BACK FROM THE BRINK

MAMMALS

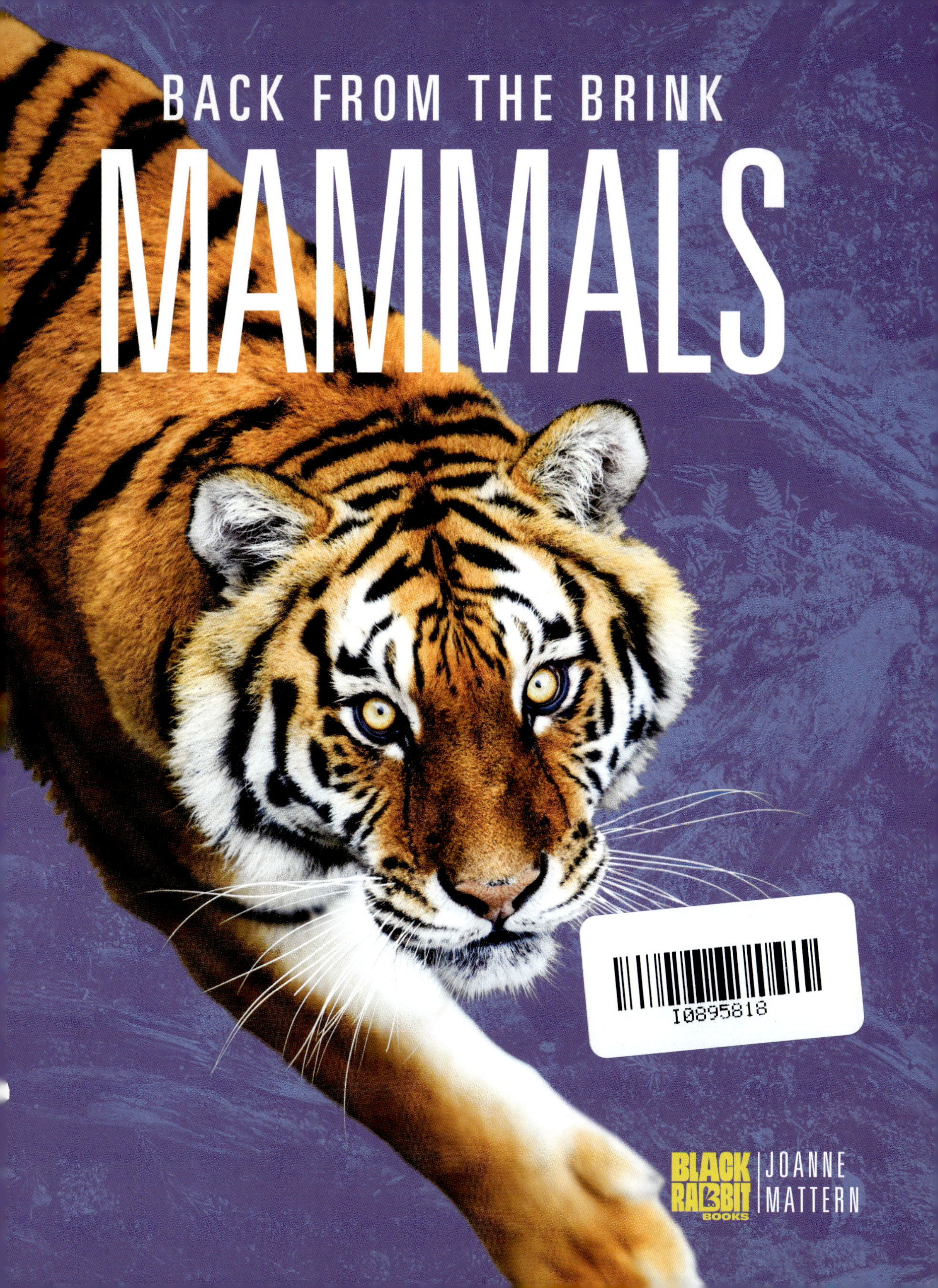

BLACK RABBIT BOOKS

JOANNE MATTERN

TABLE OF CONTENTS

1

American Bison

Until the 1800s, millions of American bison lived on the Great Plains. American Indians hunted the bison. They ate the meat. They used the bones to make tools. They made clothing out of the thick, furry hides.

Then white settlers came. They killed almost all the bison. They took away the grasslands where bison lived. By 1889, there were about 500 bison left.

The US government and Native nations stepped in to help. They set up safe places for the bison. Slowly, these big animals came back. Today, about 31,000 wild bison roam the country.

2

Florida Panther

These big cats once lived all over Florida. But too many people hunted them. People also built houses on the panther's land. Many cats were hit by cars.

Scientists jumped in to help. They built special bridges to help the cats cross the road. The government set aside land. This was a safe place for the cats to live. Today, about 200 panthers live in Florida. People hope that number will grow even more.

Did You Know?
March 16 is Save the Florida Panther Day.

3 Black-Footed Ferret

These animals live in grasslands in North America. They eat mainly prairie dogs. In the 1960s, towns were built in the grasslands. People poisoned the prairie dogs. That meant the ferrets died too.

People thought all the ferrets were gone. Then in 1981, a dog found some on a ranch in Wyoming. The government hurried to save the ferrets. Zoos raised them. Then they were let go in safe places. Today, the ferret still faces danger. But it has a good chance to survive.

Think About It
How can the government and zoos work together to save animals?

4 Giant Panda

The giant panda faced a lot of problems. Pandas only eat bamboo. But many bamboo forests were cut down. The panda had no food and no place to live. By the 1960s, the animals were in big trouble.

The Chinese government stepped in to help. They made it illegal to cut down bamboo or hunt pandas. They sent pandas to zoos all over the world to breed.

As of 2021, giant pandas are no longer endangered. They still face many problems. But China and zoos are helping to keep this cuddly animal alive.

Did You Know?
Giant pandas usually have one cub every two years.

5

Gray Wolf

Many people hate or fear wolves. By the early 1900s, many wolves had been killed. The animal almost disappeared from the United States.

People wanted to save the gray wolf. The government passed a law to stop hunting wolves. This protected the animals. But more needed to be done. In 1982, gray wolves were let go in some national parks. Today, the gray wolf is doing well in the United States.

Think About It

How did the return of gray wolves help the areas where it lives?

6

Tiger

Tigers were once found all over Asia. Now, their habitat is disappearing. Many are hunted for their fur. Others are killed to make medicine. It seems that tigers are doomed to become extinct.

Several countries have laws to protect tigers. Governments set up safe places for them to live. Zoos are also helping to save tigers all over the world. These animals breed well in captivity.

Did You Know?

There are more tigers living in zoos than in the wild.

MORE TO EXPLORE

WHERE IN THE WORLD?

AMERICAN BISON
Great Plains of North America

BLACK-FOOTED FERRET
Great Plains of North America

FLORIDA PANTHER
Florida, United States

GRAY WOLF
Northern parts of North America, Europe, and Asia

TIGER
India and southeast Asia

GIANT PANDA
Central China

MORE TO EXPLORE

FANTASTIC FACTS

Despite their huge size, bison can run faster than horses.

One black-footed ferret can eat more than 100 prairie dogs in a year.

A Florida panther's favorite foods are deer and wild hogs.

China owns all the giant pandas in the world.

All dog breeds are related to wolves.

Unlike most cats, tigers love water and are great swimmers.

MORE TO EXPLORE

COOL COMPARISONS

These animals are making a comeback.
How many live in the wild today?

American Bison
about 20,000

Black-Footed Ferret
about 350

Florida Panther
about 200

Giant Panda
about 1,800

Gray Wolf
about 6,000

Tiger
about 4,000

MORE TO EXPLORE

RESOURCES

Glossary

breed (BREED) To give birth to or produce offspring.

captivity (kap-TIV-i-tee) The state of being kept in a place and not being able to leave or be free.

extinct (ik-STINGKT) When a plant or animal species no longer exists.

habitat (HAB-uh-tat) The place where a plant or animal grows or lives.

hide (HIDE) The skin of a large animal, such as a cow or a buffalo.

illegal (ih-LEE-guhl) Something that is against the law or rules.

poison (POI-zuhn) Something that can kill or hurt people or animals when swallowed.

Read More

Gagne, Tammy. *Endangered Mammals.* San Diego, CA: BrightPoint Press, 2024.

Golkar, Golriz. *Endangered Mammals Around the World.* North Mankato, MN: Pebble, 2025.

Index

TOP RANK is published by Black Rabbit Books, P.O. Box 227, Mankato, MN, 56002.

• Top Rank is an imprint of Black Rabbit Books • Edited by Ana Brauer • Book designed by Jason Knudson • Series designed by Danny Nanos • Photographs © Alamy Stock Photo/Larry Richardson / Danita Delimont, 8, Laura Romin & Larry Dalton, 11; Dreamstime/Pavel Shlykov, 14, 21, 23, Timothy Mimm, 10, 21, 23, Ylivdesign, 4, 21, 23; Freepik/goinyk, 12, Josef Svoboda, 6–7, 21, 23; Minden Pictures/Gerard Lacz / FLPA, 20; Shutterstock/dangdumrong, 13, Ernie, 17, 21, 23, Jean-Edouard Rozey, 5, KensCanning, 15, PhotocechCZ, 18, 21, 23, Puttachat Kumkrong, 2, 19, Splendid Photos, cover, 1, The Len, cover, 1, Vaclav Sebek, 9; SuperStock/Jim Brandenburg/Minden Pictures, 16 • Printed in India

Library of Congress Cataloging-in-Publication Data: Names: Mattern, Joanne, 1963- author. | Title: Mammals : back from the brink / by Joanne Mattern. | Description: Mankato, MN: Black Rabbit Books, [2026] | Series: Back from the brink | Includes bibliographical references and index. | Ages 8–11 | Grades 2–3 | Identifiers: LCCN 2024043389 | ISBN 9781644667996 (library binding) | ISBN 9781644668313 (paperback) | ISBN 9781644668634 (ebook) | Subjects: LCSH: Rare mammals—Conservation—Juvenile literature. | Endangered species—Juvenile literature. | Classification: LCC QL706.8 .M378 2026 | DDC 599.168—dc23/eng/20250110 | LC record available at https://lccn.loc.gov/2024043389